21st Century
Basic Skills
Library

PIE GRAPHS

Favorite Colors

by Sherra G. Edgar

Cherry Lake Publishing • Ann Arbor, Michigan

2

Published in the United States of America by Cherry Lake Publishing
Ann Arbor, Michigan
www.cherrylakepublishing.com

Consultants: Janice Bradley, PhD, Mathematically Connected
Communities, New Mexico State University; Marla Conn, Read-Ability
Editorial direction and book production: Red Line Editorial

Photo Credits: Shutterstock Images, cover, 1, 16; Felix Mizioznikov/
Shutterstock Images, 4; Eric Isselee/Shutterstock Images, 6; Big Cheese
Photo/Thinkstock, 8; iStockphoto/Thinkstock, 10; Monkey Business
Images/Shutterstock Images, 12; Jupiter Images/Thinkstock, 14; Nina
Malyna/Shutterstock Images, 20

Library of Congress Cataloging-in-Publication Data
Edgar, Sherra G.
  Pie graphs / Sherra G. Edgar.
      pages cm. -- (Let's make graphs)
  Audience: K to grade 3.
  Includes bibliographical references and index.
  ISBN 978-1-62431-394-3 (hardcover) -- ISBN 978-1-62431-470-4
(paperback) -- ISBN 978-1-62431-432-2 (pdf) -- ISBN 978-1-62431-508-4
(ebook)
  1. Graphic methods--Juvenile literature. I. Title.

  QA90.E354 2013
  518'.23--dc23

                                      2013005088

Cherry Lake Publishing would like to acknowledge the work of The
Partnership for 21st Century Skills. Please visit www.p21.org for more
information.

Printed in the United States of America
Corporate Graphics Inc.
July 2013
CLFA11

TABLE OF CONTENTS

What Is a Pie Graph?

Owen wants to count his friends' pets. He can use a **graph**. Graphs show **data**.

My Friends' Pets

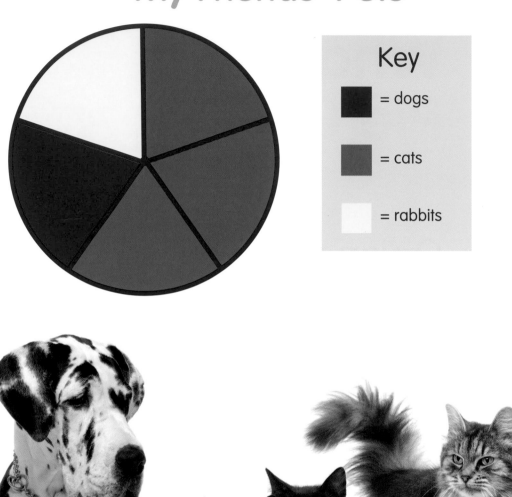

Key
■ = dogs

■ = cats

□ = rabbits

Owen made a **pie graph**. Pie graphs show parts of a whole. A pie graph is a circle.

Favorite Summer Fun

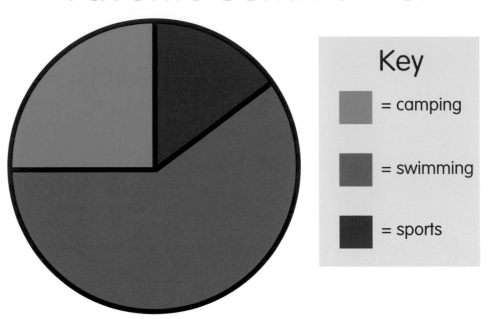

Key

■ = camping

■ = swimming

■ = sports

Inside the circle are slices.
Each slice stands for an
amount of data.

Do You Like Pizza?

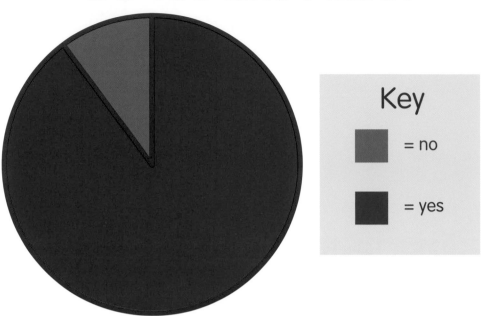

Key

■ = no

■ = yes

The slices are different colors. The **key** tells what each color stands for. Some slices are small. Some slices are big.

Making a Pie Graph

Sara and her five friends like snacks. What are their favorites?

Our Favorite Snacks

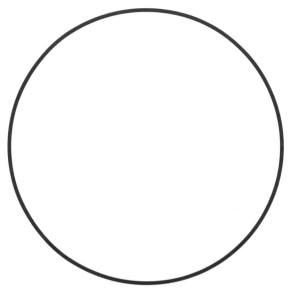

Sara made a pie graph to show the snacks. She wrote the title "Our Favorite Snacks." She made a circle.

Our Favorite Snacks

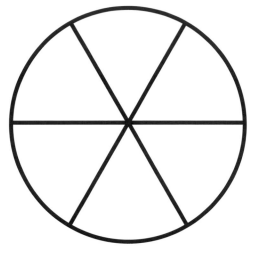

Sara drew lines to show six slices. There is one slice for each person.

Our Favorite Snacks

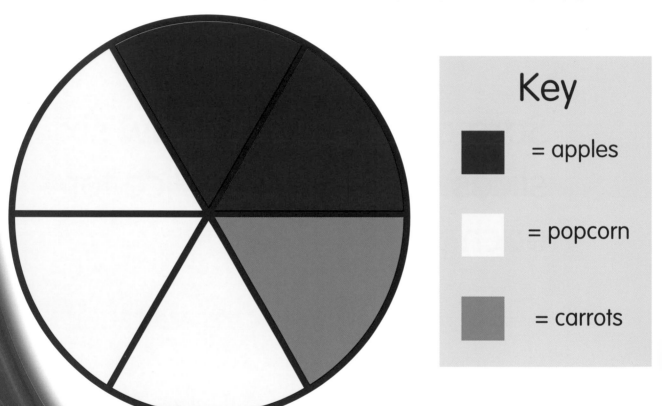

Key

■ = apples

□ = popcorn

■ = carrots

Sara used colors to show the snacks. Which snack do Sara's friends like best? Sara's friends like popcorn.

You Try It!

What are your friends' favorite colors? Make a pie graph. Draw one slice for each person. Color in each slice.

Find Out More

BOOK

Piddock, Claire. *Line, Bar and Circle Graphs.* New York: Crabtree, 2010.

WEB SITE

Mr. Nussbaum Ice Cream Graphing
http://mrnussbaum.com/graphing/
Practice reading and making graphs about ice cream.

Glossary

amount (uh-MOUNT) how many or how much there is of something

data (DEY-tah) amounts from a graph

graph (GRAF) a picture that compares two or more amounts

key (KEE) a table that tells what each color means on a pie graph

pie graph (pahy graf) a circle graph that shows parts of a whole

Home and School Connection

Use this list of words from the book to help your child become a better reader. Word games and writing activities can help beginning readers reinforce literacy skills.

a	favorite	of	snacks
amount	favorites	one	stands
an	five	our	tells
and	for	Owen	the
are	friends	parts	their
best	graph	person	there
big	graphs	pets	title
can	he	pie	to
circle	her	popcorn	use
color	his	Sara	used
colors	in	she	wants
count	inside	some	what
data	is	show	which
do	key	six	whole
different	like	slice	wrote
draw	lines	slices	your
drew	made	small	
each	make	snack	

Index

About the Author

Sherra G. Edgar is a former primary school teacher who now writes books for children. She also writes a blog for women. She lives in Texas with her husband and son. She loves reading, writing, and spending time with friends and family.